From Flowers To Foliage

the long awaited sequel to a procession of flowers (at least one person waited about 20 minutes)

By Becky Bereman Grimes

ISBN 1460942264

Published by Alternatives Unlimited
324 S. Brooks St.
Sheridan, WY 82801

Thanks Charlie!

Also by Becky Bereman Grimes

Leaf Art (lefart):
A Joyful and Playful Look at Leaves...and some poems
Leaf Art ABC...and some dances
Larry or Lariet, an adventure of a little buffalo calf
Destination: Mammoth Hot Springs for a Very Special Event
Destination: Grand Canyon of the Yellowstone
A Procession of Flowers
From Flowers to Foliage
Bird Tails

Giddy-Up!
An original 2010 Leaf Art (lefart) creation by
Becky Bereman Grimes
au@vcn.com

A Beautiful September Day
September 4, 2010

In his element plying his trade.

We are delighted to see
the American Dipper is
plying its trade as well.

This is the third
or fourth cutting
for many of
these fields.

It looks like
there is
more to
come.

Where are the Fish and Cows?
September 11, 2010

This photo is some what reminiscent of a famous Charles Belden photo of cows trailing home in a blizzard.

In September the lone tree stands
tall and green over changing
colors of willow.

It's a gorgeous day but where are the fish? There's just a couple of cows and some blue winged teal.

A Glorious Day
September 17, 2010

Loved For A Lifetime.
He carves our initials in
the bark of an aspen.

It's the kind of fall day that makes you want to sing. The breeze in the aspen does just that!

The sky so blue, the greens so green and
the reds coming on.

We can see far
down into the
canyon of the
Tongue River.

Steamboat stands
guard over this
stretch of the planet.

There must be springs on that side of Steamboat. There are many springs on this side. It's gentler over here though.

Portraits of Dead-eye Charlie and his sidekick Black Bart.

They don't seem too spooky even though
it is hunting season. Steamboat is
framed nicely between pine and aspen.

River of Gold

September 18, 2010

Red aspen again!

How do these turn red when they are so
near to this stand of sizzling gold?

Sizzle indeed!

These trees are hot.

The trees on the
way over the
hill are just as
spicy! The gold
spills out onto
the ground.

How interesting that this stand has such a
variety of color. The willows along the creek
are spectacular and this cow knows it!

WOW!!!

The sunlight
spotlights this tall
grass in the willows.

This grove and
the grove at the
rock smile take
my breath away!

This beauty to
behold bids us
farewell for today!

Caution: Slow Fish
September 26, 2010

It's a slow day at the fish market but the scenery is nice. The Garden of the Gods looks down upon us.

As we approached the lone tree appeared more beautiful and the leaves were so bright.

I would say this is just as beautiful as the fall in upstate New York! The backlit willows glimmer and twinkle in the breeze.

We have seen signs that a beaver has been working on this dam all summer.

He is happiest
when he is fishing
and so am I!

Over his head is
the lone tree
aflame on one
side!

23

Is there a more beautiful sound than water moving over rocks?

A Procession of Willows

June 5, 2010

June 26, 2010

October 20, 2010

Sheridan County Vista

The Wily Grouse Again

September 27, 2010

Look closely
and you will
see a moose and
her calf.

A very nice place
for the cow and
calf to make their
home.

This aspen is
giving the moose
and its calf a
standing 'O'!

The aspen are lush and luminous! This corridor of trees look like they are in a park or some fine estate.

This dappled
light invites
us to take a
walk along a
trickling
stream...which
leads us to
another
dappled
grove. The
path begs to
be walked
through
slowly...

...listening to dry
leaves crunch, hearing
the sighing of the
wind in the branches

and the trickle of the stream and smelling the rich loamy earth. The grove floor is beautiful with fallen golden leaves, and creeping maple and strawberry leaves turning to t
heir fall colors.

These are the
tallest aspen I
have ever
seen!

For a scale
you can see
Charlie in
the
photo below.

Grouse Again
October 1, 2010

We finally saw a blue grouse.

The bushes are a pretty rusty color. We can see changes in the landscape below, and it won't be long until winter comes to
this corner of the world.

He finally gets his grouse...a dusky blue.
The sun is setting earlier and I catch the
sunlight lighting up a stand of aspen.

Steamboat, looks like the prow of an
ocean liner making its way in the setting
sun.

Will Grousing
October 20,2010

Charlie's brother, Will, came grousing with us and loved the walk as much as we did. The storms are getting colder and more often but it is now late October and that is to be expected. This little waterfall is beginning to build up ice.

I hope you will enjoy this review of the summer and Fall in the Big Horn Mountains. The prequel to this book is

A Procession of Flowers.

In it you will find stunning photos like this.

www.ingramcontent.com/pod-product-compliance
Lightning Source LLC
Chambersburg PA
CBHW050845290526
45792CB00002B/523